THE GREEN CURE

The Science and Magic of Indoor Plants for a Happier, Healthier You

By

Bert D. Bouffard

Intentionally left blank

Copyright ©Bert D. Bouffard, 2024

All rights reserved. No part of this publication may be reproduced, distributed, or transmitted in any form or by any means, including photocopying, recording, or other electronic or mechanical methods without the prior written permission of the copyright owner, except in the case of brief quotations embodied in critical reviews and certain other non commercial use permitted by copyright law.

Table of Contents

INTRODUCTION   5

CHAPTER 1   9
A World of Green Awaits   9

CHAPTER 2   19
Breathing Easy: Indoor Plants as Natural Air Purifiers   19

CHAPTER 3   28
Cultivating Calm: The Mental Health Benefits of Indoor Plants   28

CHAPTER 4   37
Blooming Productivity: How Plants Enhance Your Work Life   37

CHAPTER 5   47
Restful Nights: Indoor Plants in Pursuit of Better Sleep   47

CHAPTER 6   57

The Perfect Match: Choosing the Right Indoor Plant for You   57

CHAPTER 7   67

Plant Parenthood 101: Essential Care for Indoor Plants   67

CHAPTER 8   77

Troubleshooting Tips: Keeping Your Indoor Plants Thriving   77

CHAPTER 9   86

Beyond the Basics: Styling Your Home with Indoor Plants   86

CHAPTER 10   95

The Green Revolution: Indoor Plants for a Sustainable Future   95

CONCLUSION   102

APPENDIX   110

Intentionally left blank

INTRODUCTION

Occasionally, do you get the impression that the entire world is weighing down on you? The incessant stress of work, the hustle of city life, and the ever-present light of screens are all that are necessary to make everyone wish for a moment of peace and a breath of fresh air. What would happen, however, if that refuge of serenity could be found right in the comfort of your own home?

The book "The Green Cure: The Science and Magic of Indoor Plants for a Happier, Healthier You" urges you to integrate the transforming power of nature into your living environment by exposing you to the science and magic of indoor plants. There is more to indoor plants than merely being beautiful components; they are quiet companions in our attempts to create and sustain a state of well-being. Put yourself in a world where:

Houseplants fulfil the job of natural air purifiers, eliminating toxins and pollutants from the air and leaving behind a clean and revitalising atmosphere. This results in the air that you breathe being cleaner.

Simply being in the company of plants has been demonstrated to greatly lessen stress, anxiety, and even depression, according to a number of studies. Picture yourself in a location where negative feelings are replaced by a sense of calm and tranquilly.

Concentration improves, and productivity skyrockets: Struggling to concentrate at your home office? Research reveals that putting plants into your workspace may boost attention and enhance cognitive function, leading to greater productivity and a sense of success.

Restful sleep becomes a reality: Certain plants emit pleasant scents that stimulate relaxation and deeper sleep, allowing you to wake up feeling refreshed and ready to confront the day.

But the benefits don't stop there. This book will be your entire introduction to the exciting world of indoor plants. You'll look into the intriguing history of these green pals, from their presence in ancient civilizations to their rise in modern home design. We'll study the science driving their extraordinary skills, revealing the research that reveals their effect on our physical and mental well-being.

But this isn't merely a book about the "why"; it's also about the "how." We'll offer you with the information and skills to properly cultivate your own indoor hideaway. You'll learn how to choose the ideal plants for your needs and lifestyle, from the low-maintenance snake plant to the air-purifying peace lily.

"The Green Cure" will offer you with:

Essential care tips: Discover the basics of adequate watering, lighting, and nutrients to ensure your plants develop.

Troubleshooting techniques: Learn how to detect and address common plant ailments, from pests and diseases to wilting leaves.

Inspiring design ideas: Unleash your inner interior designer and learn how to create a beautiful and practical living room bursting with vivid plants.

Whether you're a seasoned gardener with a passion for plants or a complete beginner hoping to add a touch of nature to your home, "The Green Cure" will be your guide on your road towards a happier, healthier you. So, detach, relax, and connect with the calming effect of nature. Let the green cure begin!

CHAPTER 1

A World of Green Awaits

In our fast-paced, technology-driven society, it's easy to feel alienated from nature. We spend endless hours inside, bathed in artificial light, inhaling recycled air. But what if there was a way to bring a bit of the natural world back into our homes, a way to create a refuge of calm and quiet among the daily hustle? Enter the magical domain of indoor plants.

Beyond Aesthetics: The Hidden Power of Indoor Plants

For decades, indoor plants have adorned our homes and public places, their brilliant colors and textures providing a touch of life and beauty to our surrounds. However, their importance goes well beyond aesthetics. These quiet friends carry a secret power, a capacity to greatly enhance our physical and emotional well-being.

Imagine entering into your living room after a hard day, the air cleaned and revitalised by the presence of fresh greens. Picture yourself working in your home office, surrounded by plants that not only enrich the décor but also sharpen your attention and raise your productivity. Envision a pleasant night's sleep calmed by the relaxing smells pouring from your bedroom plants.

This isn't a fiction — it's the reality that indoor plants may generate. Scientific study has started to disclose the astonishing potential of these living air purifiers, stress reducers, and mood boosters. Studies have demonstrated that indoor plants can:

Purify the air: Plants work as natural filters, absorbing toxins like formaldehyde, benzene, and trichloroethylene, typical pollutants present in our homes from cleaning products, furniture, and electronics. A NASA clean air study undertaken in the 1980s revealed the efficiency of some indoor plants in eliminating these pollutants, opening the path for additional research into their air-purifying qualities.

Reduce tension and anxiety: Studies show that merely being among plants may have a soothing impact on the nervous system. The act of caring for plants may also be therapeutic, bringing a feeling of purpose and achievement.

Improve mood and well-being: Exposure to nature has been demonstrated to have a good influence on mood, lowering symptoms of sadness and anxiety. Indoor plants may bring a bit of the outside in, creating a connection to nature that can raise the soul and promote general well-being.

Boost productivity and focus: Research shows that bringing plants into the workstation might boost cognitive function and concentration. This may lead to improved productivity and a feeling of success in the job.

Promote better sleep: Certain plants like lavender and jasmine generate relaxing smells that may promote relaxation and deeper sleep.

These are just a handful of the well-documented advantages of growing indoor plants. As study continues, we may uncover many more ways these green friends help to our health and pleasure.

A Journey Through Time: The History of Indoor Plants

Our obsession with indoor plants is not a new phenomena. Humans have been introducing plants into their living environments for millennia, motivated not merely by aesthetics but also by a desire to connect with nature and exploit its advantages.

Ancient Civilizations: Evidence reveals that the ancient Egyptians, Greeks, and Romans all utilised plants to beautify their temples, palaces, and houses. Egyptians integrated papyrus reeds and palm trees into their houses, while the Greeks and Romans thought plants carried symbolic value and employed them in religious rites and for ornamental reasons.

The Middle Ages: During this era, monasteries and castles commonly had inside gardens, acting not only as a source of beauty but also as a method of distributing herbs and medicinal plants. The contained nature of these places made the growing of indoor plants a requirement, spurring the development of systems for sustaining greenery in regulated surroundings.

The Renaissance and Beyond: The Renaissance era witnessed a revived interest in classical learning and culture, leading to a renaissance of the use of plants in interior design. Wealthy Europeans started collecting exotic plants from throughout the globe, presenting them in their houses and gardens. This inspired a rising passion with horticulture, leading to the development of novel ways for producing and caring for a greater range of indoor plants.

The 20th Century and Onwards: The 20th century witnessed a change in the way indoor plants were seen. With developments in technology and mass manufacturing, indoor plants became more accessible and inexpensive, leading to a spike in their appeal. They were no longer simply the realm of the rich but could be enjoyed by individuals from all walks of life. Today, indoor plants are a hallmark of contemporary home design, with various types available to suit every taste and style.

The Evolving Relationship with Indoor Plants

The growth in popularity of indoor plants in the 20th century wasn't only about price. It corresponded with a rising awareness of the harmful implications of contemporary life. The trend towards urbanization and the growing dependence on artificial materials in building led to a craving for a connection with nature. Indoor plants became a method to bring a little of the outside in, creating a feeling of peace and a reminder of the natural world among the concrete jungle.

This tendency persisted into the 21st century, fuelled by a growing amount of studies on the advantages of biophilia — the innate human connection with nature. As studies showed the good influence of plants on air quality, stress levels, and general well-being, indoor plants became more than simply ornamental components. They came to be viewed as key contributions to establishing healthy and pleasant living conditions.

The Rise of Indoor Plant Enthusiasts

The internet has played a major role in the current growth of interest in indoor plants. Social media sites like Instagram and Pinterest are replete with breathtaking photographs of lush indoor rainforests and perfectly kept plant collections. Online groups have grown up, linking plant enthusiasts worldwide, establishing a feeling of shared interest and giving a forum for study, guidance, and inspiration.

These internet networks have also democratized access to knowledge and resources. Beginners may discover thorough care recommendations for individual plants, while experienced plant parents can discuss ideas and strategies for developing successful indoor gardens. This greater accessibility has empowered a new generation of indoor plant aficionados, making it simpler than ever before to grow a tiny bit of nature inside their homes.

A Look Towards the Future

As we go ahead, the future of indoor plants seems bright. Technological improvements, such as self-watering pots and smart grow lights, are making it even simpler to care for plants, widening their appeal to individuals with hectic lives. Additionally, the increased awareness of sustainability is leading to an emphasis on eco-friendly gardening methods. This includes utilising organic fertilizers, picking plants local to your location, and opting for repurposed or upcycled containers.

The future of indoor plants is likely to see a continuous emphasis on their health advantages. With increased focus on mental well-being and stress reduction, architects and designers may incorporate indoor plants into living and offices as a matter of course. Furthermore, study into the individual air-purifying capabilities of various plants might lead to the creation of specialised solutions for enhancing indoor air quality in certain circumstances.

A World of Green Awaits

Whether you're a seasoned gardener or a complete beginner, there's a place for indoor plants in your life. They offer a wealth of benefits, from purifying the air you breathe to reducing stress and anxiety. They may boost your concentrate, improve your sleep, and even raise your mood. They are living works of art, providing a touch of life and beauty to your surrounds. The world of indoor plants beckons, giving a route towards a happier, healthier self.

This chapter has studied the history of indoor plants, their changing significance in our lives, and the exciting possibilities they promise for the future. In the coming chapters, we'll delve deeper into the science behind their benefits, explore the vast array of indoor plants available, and equip you with the knowledge and tools you need to cultivate your own thriving indoor oasis. So, turn the page and embark on this journey with us. Let the green cure begin!

CHAPTER 2

Breathing Easy: Indoor Plants as Natural Air Purifiers

The air we breathe is crucial for life, however the quality of indoor air may be unexpectedly bad. Our homes and workplaces may host a myriad of contaminants, including volatile organic compounds (VOCs) emitted by construction materials, furnishings, cleaning products, and personal care items. These pollutants may irritate the eyes, nose, and throat, and have been related to many health concerns, including headaches, dizziness, and respiratory disorders.

Fortunately, there's a natural remedy easily available: indoor plants. These green miracles work as living air purifiers, collecting toxins and pollutants from the air, leaving you with a cleaner, healthier atmosphere to breathe in.

Unveiling the Toxin-Fighting Power of Plants

Plants have a unique capacity to take in carbon dioxide and release oxygen via the process of photosynthesis. However, their air-purifying properties go beyond this fundamental function. Plants have microscopic holes on their leaves called stomata, which they utilise to collect water and carbon dioxide. These stomata also operate as filters, taking in air from the surrounding environment. As the air travels through the stomata, certain pollutants attach to the wet surfaces of the leaves, while others are absorbed by the plant and broken down into harmless compounds.

The particular air-purifying capacities of a plant rely on various aspects, including the species, the size of the plant, and the quantity of pollutants present in the environment. However, research has found many common indoor plants that are especially excellent in removing toxins from the air. These include:

Snake Plant (Sansevieria Trifasciata): A low-maintenance champion, the snake plant not only eliminates typical VOCs like formaldehyde and benzene but also releases oxygen at night, making it a wonderful option for bedrooms.
Spider Plant (Chlorophytum comosum): Easy to maintain for and famed for its cascading leaves, the spider plant tackles a variety of poisons, including formaldehyde, xylene, and carbon monoxide.
Peace Lily (Spathiphyllum wallisii): This attractive plant thrives in moderate light and humidity and is efficient in removing benzene, ammonia, and trichloroethylene.
Dracaena (Dracaena spp.): These flexible plants come in varied sizes, making them suited for

diverse areas. They are known to remove formaldehyde, xylene, and benzene.

Golden Pothos (Epipremnum aureum): Another low-maintenance choice, the golden pothos handles formaldehyde, toluene, and xylene.

NASA's Clean Air Study: Plants to the Rescue

In the 1980s, NASA performed a series of clean air experiments to study methods to enhance air quality for astronauts in cramped space stations. These research, headed by Dr. B.C. Wolverton, revealed the efficiency of various indoor plants in eliminating common VOCs from sealed rooms. While the research design had limits – the quantity of plants employed and the size of the chambers don't immediately transfer to ordinary household conditions – it gave vital insights into the air-purifying capacity of indoor plants.

The study identified many plants, including the snake plant, spider plant, and peace lily, as especially good in removing formaldehyde, benzene, and trichloroethylene - all typical pollutants present in indoor environments. While the NASA research didn't measure the precise quantity of pollutants eliminated in a normal residential setting, it spurred broad interest in the air-purifying qualities of indoor plants.

Creating a Healthier Indoor Environment

So, how can you harness the power of indoor plants to create a healthier indoor environment? Here are some crucial considerations:

Choosing the Right Plants: While many indoor plants provide air-purifying advantages, some are more effective than others at eliminating certain contaminants. Consider the prevalent pollutants in your house and pick plants that target those chemicals. Research online resources or check with a local nursery to select the finest plants for your requirements.

Quantity Matters: The amount of plants you require depends on the size of your area and the degree of pollution. As a general guideline, NASA recommended having at least one medium-sized plant (8-10 inches in diameter) every 100 square feet of floor area.

Placement is Key: Strategically put your plants around your house. Pollutants tend to gather around furniture, electronics, and cleaning supplies. Positioning plants in these regions may increase their efficacy.

Maintaining Plant Health: Healthy plants are more effective air purifiers. Ensure your plants get appropriate sunshine, water, and fertilizer to perform at their optimum.

Beyond Air Purification: Additional Benefits of Indoor Plants

While air cleaning is a big advantage of indoor plants, it's not the only one. Plants also:

Increase Humidity: Many plants emit moisture into the air, which may assist to battle dry indoor air, particularly during winter months. This may help dry skin, sore throats, and respiratory difficulties.

Reduce Stress and Improve Mood: Studies have shown that spending time near plants may have a soothing impact on the neurological system. The act of caring for plants may also be soothing, offering a feeling of achievement and building a connection with nature.

Boost Productivity: Research shows that bringing plants into the workstation may boost cognitive function and focus, leading to higher productivity and a feeling of well-being.

Creating a Living Air Filter: Combining Plants for Maximum Impact

While individual plants offer air-purifying benefits, boosting their effect may be performed by building a living air filter. This comprises clustering multiple plants together in close proximity. As the plants transpire, they establish a microclimate with increased humidity, which further improves their potential to trap and absorb pollutants.

Here are some suggestions for building a live air filter:

Variety is Key: Choose plants with varied heights and leaf kinds to form a visually interesting arrangement.

Consider Light Requirements: Group plants with identical light needs to assure they all grow.

Utilize Containers with Drainage: Ensure appropriate drainage to avoid root rot. Trays loaded with stones and water may help retain humidity.

Rotate Plants Regularly: This fosters uniform development and enables all the plants in the group to benefit from optimum ventilation and light exposure.

A Breath of Fresh Air: The Future of Indoor Plant Air Purification

The future of indoor plant air purification presents fascinating potential. Researchers are studying strategies to maximise the air-purifying capacities of individual plants by adjusting growth circumstances or even genetically manipulating them. Additionally, the incorporation of plants into architecture design is gaining popularity. Architects and designers are adding living walls and green roofs into projects, not just for aesthetic reasons but also to enhance interior air quality and create a better living and working environment.

Conclusion: Breathe Easy with Indoor Plants

Indoor air quality is an increasing problem, but the remedy may be as easy as adding a bit of green to your surroundings. Indoor plants are not simply beautiful components; they are natural air purifiers, capable of eliminating dangerous poisons and pollutants from the air we breathe. By picking the correct plants, arranging them strategically, and building living air filters, you can create a cleaner, healthier atmosphere for yourself and your loved ones. So, take a big intake of fresh, plant-filtered air and appreciate the myriad advantages that indoor greenery has to offer.

CHAPTER 3

Cultivating Calm: The Mental Health Benefits of Indoor Plants

In our fast-paced, technology-driven society, stress and anxiety are widespread. We manage busy job schedules, handle complicated relationships, and confront a continual assault of information overload. This unrelenting pressure may take a toll on our mental well-being, leaving us feeling overwhelmed, fatigued, and alienated. But despite this modern-day tumult, an unexpected source of peace may be found: indoor plants.

From Stress to Serenity: How Plants Boost Your Mood

Studies have shown that merely being among plants may have a tremendous influence on our mental state. The act of seeing nature, even in the form of indoor greenery, may activate a relaxation response in the body, reducing stress chemicals like cortisol and encouraging emotions of peace and contentment.

Biophilia and the Connection with Nature: Our intrinsic connection with nature, known as biophilia, plays a part in this soothing impact. Studies show that exposure to natural components, especially plants, may stimulate the parasympathetic nervous system, responsible for our rest-and-digest response. In contrast, the sympathetic nervous system, responsible for the fight-or-flight response, becomes less active in the presence of plants.

Reduced Stress and Anxiety: Research has demonstrated the stress-reducing advantages of engaging with plants. Studies have indicated that observing or caring for plants may reduce blood pressure, calm heart rate, and decrease levels of the stress hormone cortisol. Additionally, the process of fostering these live creatures may give a feeling of control and purpose, giving a pleasant escape from everyday problems.

Improved Mood and Well-being: Beyond stress reduction, plants may have a good influence on mood and well-being. Studies show that exposure to greenery might promote emotions of happiness, minimise symptoms of sadness, and enhance general emotional well-being. The rich colors and textures of plants may be visually engaging and uplifting, while the act of cultivating them creates a feeling of success Lowering

Lowering the Pressure: Plants and Cardiovascular Health

The stress-reducing benefits of indoor plants may have a good influence on cardiovascular health. Chronic stress is an established risk factor for high blood pressure and heart disease. By decreasing stress levels, plants may help to a healthy cardiovascular system.

Reduced Blood Pressure: Studies have shown that spending time with plants or even seeing environmental landscapes might contribute to a drop in blood pressure. This is likely owing to the stress-reducing action of plants, which may relax blood vessels and enhance blood flow.

Improved Heart Rate Variability: Heart rate variability (HRV) refers to the natural fluctuation in the time between heartbeats. Higher HRV is related with improved cardiovascular health and resistance to stress. Some research show that exposure to nature, including indoor plants, might enhance HRV, perhaps leading to a healthier heart.

Promoting Relaxation: The soothing influence of plants may encourage relaxation and improved sleep, both of which are vital for cardiovascular health. Chronic lack of sleep and stress may lead to high blood pressure and other cardiovascular disorders.

Fostering Focus: The Link Between Plants and Mental Clarity

In our information-saturated environment, keeping attention and concentration may be a problem. However, research shows that adding plants into your workstation might enhance cognitive performance and promote productivity.

Enhanced Attention: Studies have shown that exposure to plants may boost attention span and minimise mental weariness. The presence of flora may aid to filter out distractions and provide a more calm working atmosphere.

Increased Memory: Research shows that witnessing natural sceneries, including indoor plants, may boost short-term memory and cognitive performance. This may be especially advantageous for students and professionals who need to remember knowledge and execute complicated activities.

Reduced Cognitive weariness: Spending time near plants may help to reduce mental weariness, leading to greater alertness and attention throughout the day. This is particularly crucial in professional contexts where long hours and difficult activities may lead to mental weariness.

Creating a Sanctuary of Calm: Tips for Using Plants to Enhance Mental Well-being

Here are some strategies for integrating plants into your life to optimise their mental health benefits:

Place Plants Strategically: Position plants in rooms where you spend the most time, such as your living room, bedroom, or home office.

Choose Plants with Soothing Scents: Certain plants, including lavender and chamomile, generate relaxing smells that may further induce relaxation and increase sleep quality.

Create a Living Wall: A vertical garden in your house may be a visually appealing method to include a considerable quantity of greenery and boost the relaxing impact.

Connect with Nature Through Care: Take time to care your plants, watering them, misting them, and repotting them when required. The act of caring for live creatures may be a grounding and soothing experience.

Embrace the Creative Process: Use plants to create a tranquil and stimulating workstation. Experiment with various groupings, pot colors, and textures to discover a design that calls to you.

Beyond Science: The Magic of Indoor Plants

The advantages of indoor plants extend beyond the field of scientific study. There's a particular enchantment linked with these live partners. Caring for plants provides a feeling of duty and care, building a connection with nature that may be extremely satisfying.

Plants may also act as a source of inspiration and creativity. Their brilliant colors, complicated forms, and continual development may stimulate new thoughts and increase problem-solving abilities. Simply witnessing the peaceful persistence of plants may bring a feeling of perspective and tranquilly during hard times.

Cultivating a Greener Mind and Body

Indoor plants are great friends in our quest for a happier, healthier mind and body. Their capacity to decrease stress, boost mood, induce relaxation, and even increase cognitive function makes them great tools for negotiating the complexity of contemporary life. By adding plants into our homes and offices, we can create sanctuaries of quiet that foster our mental well-being and promote a feeling of tranquilly among the daily scurry.

In the following chapter, we'll investigate how indoor plants might increase your productivity and attention in the office. We'll look into studies on the advantages of plants in workstations and give practical strategies for building a healthy and productive green refuge.

CHAPTER 4

Blooming Productivity: How Plants Enhance Your Work Life

The contemporary office may be a fertile ground for stress and distraction. Long hours, tough deadlines, and the continuous buzz of technology may leave us feeling overwhelmed and unproductive. But what if there was a simple, natural approach to raise attention, enhance creativity, and improve your general well-being in the office? Enter the world of indoor plants.

Sharper Minds, Better Results: Plants and Workplace Efficiency

Research reveals that adding plants into your desk may dramatically boost your productivity and efficiency. Studies have demonstrated that the presence of greenery may boost cognitive performance, decrease stress, and create a more cheerful and inspirational work atmosphere.

Improved Attention and Focus: Studies have shown that exposure to plants may boost attention span and minimise mental weariness. The soothing influence of plants may assist to filter out distractions and create a more quiet atmosphere, enabling you to concentrate on the work at hand.

Enhanced Cognitive performance: Research shows that witnessing greenery might boost short-term memory and cognitive performance. This is especially important for knowledge workers who need to remember information and accomplish difficult tasks.

Increased Creativity: Studies studying the relationship between nature and creativity reveal that exposure to plants might ignite new ideas and boost problem-solving abilities. The organic forms, textures, and brilliant colors of plants may inspire a feeling of ingenuity and new insights.

Reduced Stress and Improved Mood: The stress-reducing advantages of indoor plants demonstrated in the preceding chapter apply straight to the workplace. By decreasing stress levels and raising mood, plants help promote a more happy and productive work atmosphere.

The University of Exeter Study: The Power of Plants in the Office

A revolutionary study undertaken by experts at the University of Exeter in the UK presents solid evidence for the favourable influence of plants on workplace productivity. The research recruited around 300 office workers who were randomly allocated to work in either an enriched setting with plants or a control environment without plants.

Increased Productivity: The research indicated that personnel working in the plant-enriched workplace saw a 15% boost in productivity compared to those in the control setting. This equates to a huge gain in productivity and efficiency.

Enhanced Creativity: The research also found that personnel in the plant-enriched workplace had a 30% boost in their problem-solving ability compared to the control group. This shows that plants may generate a more creative and imaginative work atmosphere.

Improved Well-being: Beyond productivity and creativity, the research revealed that people working with plants reported lower stress levels and better levels of well-being compared to those in the control setting. This illustrates the comprehensive advantages of introducing plants into the workplace.

Designing a Workspace for Focus and Creativity

Now that you understand the advantages of indoor plants at the workplace, let's discuss how to build a workspace that supports focus, creativity, and productivity:

Choosing the Right Plants: Opt for plants with low to moderate light needs, since many workplace spaces have minimal natural light. Consider snake plants, spider plants, or philodendrons, all noted for their simple maintenance and air-purifying characteristics.

Strategic Placement: Position plants near workstations to enhance their influence. Place them on desks, bookcases, or windowsills to generate a feeling of visual interest and boost attention.

Create a Living Wall: A vertical garden in your workspace can be a space-saving and visually stunning way to incorporate a significant amount of greenery. This can be particularly beneficial in open-plan offices, providing a sense of separation and promoting focus.

Group Plants for Impact: Creating mini living air filters by grouping several plants together can enhance their air-purifying capabilities and create a more visually appealing environment.

Let There Be Light: Maximize natural light whenever possible. Supplement with artificial grow lights if needed for plants with higher light requirements. Natural light not only benefits plants but also improves mood and alertness in office workers.

Embrace Biophilic Design: Biophilic design principles encourage incorporating elements of nature into the built environment. This can involve using natural materials like wood and stone, incorporating natural light and ventilation, and of course, adding indoor plants.

Beyond the Office: Plants for All Workspaces

The benefits of indoor plants extend beyond traditional office environments. Whether you work from home in a dedicated study, share a co-working space, or run a small business, incorporating plants can enhance your productivity and well-being.

For home offices, consider creating a designated workspace with plants to create a sense of separation from living areas and promote focus. In co-working spaces, even a small desk plant can provide a personal touch and create a more calming environment.

The Human-Plant Connection: Fostering a Greener Workplace Culture

Creating a plant-filled workplace goes beyond just adding greenery. It's about fostering a culture of shared responsibility and connection with nature. Here are some tips:

Encourage Plant Ownership: Allow employees to personalize their workspaces with small, easy-care plants. This fosters a sense of ownership and encourages a connection with the plants.

Rotate Plant Care Duties: Establish a system for rotating plant care duties. This distributes the responsibility, ensures the plants are well-maintained, and creates a sense of team spirit.

Organize Plant-Themed Activities: Host occasional workshops on plant care or team-building activities involving creating miniature terrariums or decorating pots.

Celebrate Plant Milestones: Acknowledge the growth and thriving of plants in the office. This fosters a sense of accomplishment and encourages continued care.

A Greener Future of Work: The Benefits of Biophilic Design

The benefits of incorporating plants into the workplace are gaining mainstream recognition. Leading firms globally are adopting biophilic design concepts, recognizing the beneficial influence on employee well-being, productivity, and creativity. Architects and designers are incorporating living walls, green roofs, and strategically placed indoor plants into workstation designs.

As research continues to study the relationship between nature and human performance, we should expect to see even more imaginative methods to integrate greenery into the workplace of the future.

Conclusion: Working with Nature: A Recipe for Success

Creating a workstation filled with healthy plants is not only about aesthetics; it's an investment in your well-being and productivity. By introducing plants into your workplace environment, you may encourage a feeling of concentration, creativity, and tranquilly, leading to greater performance and a more enjoyable work experience.

In the following chapter, we'll dig into the broad and bright world of indoor plants, investigating a number of solutions to meet your requirements, style, and space limits. We'll provide you with the information and skills you need to identify the appropriate plants for your indoor sanctuary and guarantee their long-term success.

CHAPTER 5

Restful Nights: Indoor Plants in Pursuit of Better Sleep

For many of us, attaining a decent night's sleep might seem like an unattainable dream. Stress, worries, and the continual glow of electronic gadgets all lead to sleep problems, leaving us feeling exhausted, unproductive, and out of sorts. But what if the cure to a pleasant night's rest could be discovered tucked in a pretty plant near your bed? Research reveals that integrating indoor plants into your bedroom might considerably enhance sleep quality.

The Science of Sleep: How Plants Promote Relaxation

Sleep is a key physiological function required for physical and mental well-being. During sleep, our bodies heal tissues, consolidate memories, and regulate hormones. When sleep is interrupted, it may lead to a cascade of negative effects, including exhaustion, poor cognitive performance, and increased risk of chronic health issues.

Indoor plants provide a natural way to supporting better sleep by addressing some of the primary elements that lead to sleep disturbances:

Stress Reduction: We have discussed the stress-reducing properties of plants in Chapter 3. By decreasing stress hormones like cortisol and fostering feelings of calm, plants may help create a more relaxed and sleep-conducive atmosphere.

Improved Air Quality: Many indoor plants work as natural air filters, eliminating toxins and pollutants from the air we breathe. This may be especially advantageous in the bedroom, where we spend a considerable chunk of our time breathing air. Improved air quality may contribute to deeper and more comfortable sleep.

Increased Humidity: Certain indoor plants emit moisture into the air, raising humidity levels. Dry air may irritate airways and impair sleep. Plants operate as natural humidifiers, producing a more pleasant resting environment.

Enhanced Sense of Well-being: The general soothing influence of plants and the connection with nature they bring contribute to a sense of serenity and well-being. Feeling calmer and more grounded before sleep might make it easier to drop off and remain asleep.

Soothing Scents: Calming Plants for a Better Night's Rest

Not all plants are made equal when it comes to encouraging sleep. Some species give extra advantages via their relaxing smells. Here are some common selections for the bedroom:

Lavender (Lavandula angustifolia): Renowned for its soothing and relaxing characteristics, lavender is a traditional option for the bedroom. Studies show that the aroma of lavender might induce relaxation and enhance sleep quality.

Snake Plant (Sansevieria Trifasciata): This low-maintenance champion not only filters out pollutants but also releases oxygen at night, further adding to a comfortable sleep environment.

Peace Lily (Spathiphyllum wallisii): This attractive plant thrives in moderate light and humidity and successfully eliminates typical air pollutants. Additionally, it creates a mild, pleasant aroma that might encourage relaxation.

English Ivy (Hedera helix): Another air-purifying powerhouse, English ivy thrives in low light and may help eliminate airborne irritants that may lead to sleep difficulties. It's crucial to remember that English ivy may be poisonous to dogs if swallowed, so keep this in mind if you have furry friends.

Valerian (Valeriana officinalis): This blooming plant has a long history of usage in herbal treatments for anxiety and insomnia. While the scientific evidence is mixed, some studies suggest that valerian may be helpful in promoting sleep. However, it's vital to contact with a healthcare practitioner before consuming valerian, particularly if you are taking any drugs.

Creating a Sleep Sanctuary with Indoor Greenery

Now that you understand the advantages of indoor plants for sleep, let's investigate how to establish a sleep refuge in your bedroom:

Choosing the Right Plants: Consider factors like light levels, humidity requirements, and maintenance needs when selecting plants for your bedroom. Opt for low-maintenance varieties that don't require frequent watering or misting. If you have pets, choose plants that are non-toxic to animals.

Strategic Placement: Don't crowd your bedroom with plants. Place them strategically to maximize their benefits. Position plants near windows for adequate light, but avoid placing them directly in your line of sight to prevent blocking airflow.

Focus on Air Quality: Choose plants known for their air-purifying capabilities and those that release moisture to improve humidity levels.

Embrace Calming Scents: For an extra dose of relaxation, incorporate plants with calming scents, like lavender or valerian (with caution, as mentioned earlier). However, be mindful of strong scents and avoid placing them directly next to your bed.

Create a Relaxing Ambiance: Combine plants with other sleep-promoting elements like soft lighting, calming colors, and comfortable bedding. This will create a holistic environment conducive to restful sleep.

Beyond Plants: Optimizing Your Sleep Environment

While indoor plants may greatly enhance sleep quality, they are simply one component of the jigsaw. Here are some extra strategies for building a sleep refuge in your bedroom:

Establish a Regular Sleep Schedule: Go to bed and wake up at predictable times, including on weekends. This helps manage your body's normal sleep-wake cycle.

Create a Relaxing Bedtime Routine: Develop a soothing ritual before bed that helps you wind down. This might involve taking a warm bath, reading a book, or practicing relaxation methods like deep breathing or meditation.

Limit Screen Time Before Bed: The blue light generated from electronic gadgets might disturb sleep patterns. Avoid using electronic gadgets for at least an hour before sleep.

Optimize Light and Temperature: Ensure your bedroom is dark, quiet, and cool for best sleep. Invest in blackout curtains, earplugs, and a thermostat to create a pleasant sleep environment.

Manage Stress: Chronic stress is a primary factor behind sleep disorders. Practice stress-management practices like yoga, meditation, or spending time in nature to decrease tension and encourage better sleep.

Exercise Regularly: Regular physical exercise might enhance sleep quality. However, avoid vigorous activity close to sleep, since it might have a stimulating impact.

Living with Nature's Rhythm: The Power of Light and Darkness

Plants play a critical part in adapting to natural light cycles. Many plants emit oxygen at night, further adding to a tranquil sleep environment. Conversely, certain plants may exhibit modest motions or produce odours throughout the day. By knowing these natural cycles, we can arrange plants strategically and create a bedroom environment that promotes our own sleep-wake cycle.

The Future of Sleep: Technology Meets Nature

The future of sleep provides intriguing prospects for further incorporating nature into the bedroom. Smart lighting systems may be tuned to replicate natural dawn and sunset, enabling a more natural sleep-wake cycle. Additionally, developments in biophilic design may lead to the incorporation of living walls or vertical gardens in beds, producing a totally immersive and peaceful sleep environment.

Nurturing Sleep with Indoor Plants

A good night's sleep is vital for our physical and mental well-being. By bringing indoor plants into your bedroom and establishing a sleep sanctuary, you may induce relaxation, enhance air quality, and create a tranquil atmosphere conducive to deep sleep. Remember, consistency is crucial. Establish a consistent sleep pattern, practice a peaceful nighttime ritual, and surround yourself with the calming influence of nature. With a little work, you may develop a sleep refuge that supports a healthy and restful night's slumber.

CHAPTER 6

The Perfect Match: Choosing the Right Indoor Plant for You

The notion of integrating colourful vegetation into your house might be exhilarating. But with the enormous selection of indoor plant alternatives available, picking the appropriate partner may seem daunting. Fear not! This chapter will provide you with the information and resources to select the best plant match for your requirements, lifestyle, and space.

Assessing Your Needs and Lifestyle

Before jumping fully into the world of houseplants, take some time to assess your lifestyle and living environment. Here are some crucial variables to consider:

Light Levels: Not all plants are made equal when it comes to light needs. Some thrive in strong, direct sunshine, while others prefer low-light circumstances. Assess the natural light levels in your house, especially the places where you want to install your plants.

Time Commitment: Be honest about how much time you can actually spend to plant maintenance. Some plants need regular watering and misting, while others are comparatively low-maintenance. Consider your schedule and pick plants that meet your time commitment.

Humidity: Certain plants demand high humidity levels, while others tolerate drier settings. Consider the usual humidity levels in your house, especially during winter months when heating systems may dry up the air.

Aesthetics: Plants come in a broad spectrum of forms, sizes, and colors. Consider your unique style and pick plants that match your décor and create the appropriate environment.

Pet-Friendly Options: If you have furry pets, it's vital to pick plants that are non-toxic to animals. Many common houseplants may be dangerous if consumed by pets.

Popular Indoor Plant Options:

Now that you have a better idea of your demands, let's investigate some popular and diverse indoor plant options:

Snake Plant (Sansevieria Trifasciata):

This low-maintenance champion is a favourite option for newbies. Snake plants withstand low light, infrequent irrigation, and dry environments. They also have air-purifying capabilities and release oxygen at night, making them excellent for bedrooms. Their upright, architectural design provides a touch of contemporary elegance to any environment.

Spider Plant (Chlorophytum comosum):

Spider plants are recognised for their flowing leaves and simplicity of maintenance. They flourish in moderate light conditions and need watering only when the dirt feels dry. These prolific growers generate baby spiderettes, enabling you to effortlessly propagate new plants and share the fun with friends and family.

Peace Lily (Spathiphyllum wallisii):

This attractive plant offers glossy green foliage and lovely white blossoms. Peace lilies like moderate, indirect light and communicate watering needs by drooping their leaves gently. They work as natural air purifiers, eliminating common home contaminants and providing a sense of refinement to any environment.

Dracaena (Dracaena spp.):

Dracaena is a genus of adaptable plants with many varieties to suit varied interests. They come in a range of sizes and leaf hues, allowing a number of aesthetic alternatives. Dracaena plants normally like moderate light and need watering when the topsoil dries up. They are recognised for their air-purifying capabilities and provide a sense of the tropics inside.

Succulents (e.g., Aloe, Echeveria, Crassula):

Succulents are a varied category of plants notable for their succulent leaves and low-water demands. They come in a breathtaking diversity of forms, sizes, and colors, giving a bit of whimsy to any area. Succulents thrive in bright, indirect light and require rarely watering, enabling even the busiest folks to enjoy the advantages of greenery.

Finding Your Perfect Green Companion

Beyond the common alternatives described above, the world of indoor plants is large and varied. Consider researching resources like online plant databases, local nurseries, and gardening books to uncover a greater range and locate the right fit for your requirements.

Here are some other suggestions for selecting your ideal green companion:

understand individual Plants: Once you have discovered a few viable possibilities, spend some time to understand their individual needs, including light requirements, watering frequency, and potential pests.

Consider Growth Habit: Think about the intended size and growth behaviour of the plant. Do you prefer a tiny plant for your desk or a towering specimen for a corner? Matching the plant's growth habit to your location can guarantee long-term success.

Think About Maintenance: Be honest about your capacity to care for plants. If you have a hectic lifestyle, choose for low-maintenance types.

Don't Be Afraid to Experiment: The charm of indoor plants resides in their variety. Start with a few plants that fit your requirements and increase your collection as you acquire expertise.

Beyond the Plant: Essential Supplies for Plant Parenthood

Every effective plant parent requires a well-stocked armoury. Here are some critical materials to guarantee the long-term survival of your indoor oasis:

Pots & Drainage: Choose pots with drainage holes to minimise waterlogging, which may lead to root rot. Match the pot size to the plant's maturity, giving for some opportunity for development. Consider utilising ornamental planters with drainage trays for a classy accent.

Potting Mix: Use a well-draining potting mix particularly intended for indoor plants. Avoid using garden soil, which might be excessively thick and contain too much moisture.

Watering Can: A watering can with a long spout provides for accurate watering, delivering water to the base of the plant and avoiding foliage.

Spray Bottle: A mister is excellent for raising humidity around plants that enjoy it, especially during cold months.

Pruning Shears: Sharp pruning shears will help you to remove dead or damaged leaves and preserve the correct form of your plants.

Houseplant Fertilizer: While not needed for all plants, a balanced houseplant fertilizer may offer critical nutrients and support healthy development, particularly for plants with lush foliage or blooming kinds. However, be careful to follow the manufacturer's directions for optimum dilution and application frequency.

Light Meter: A light meter may be a valuable tool for assessing the light levels in various regions of your house. This may be particularly important for ensuring your plants get the right light for their requirements.

Creating a Thriving Indoor Ecosystem

With the appropriate plant selection, adequate maintenance, and a little TLC, you can establish a healthy indoor environment that brings beauty, vitality, and well-being into your house. Remember, plants are living beings, and their demands may change based on the season and environmental variables. Observe your plants periodically, change their care appropriately, and don't hesitate to study remedies for any concerns that may emerge.

The Art of Plant Parenthood: Embracing the Journey

Caring for indoor plants is not only about keeping them alive; it's about creating a connection with nature and building a feeling of duty and achievement. The act of raising these living creatures may be a source of pleasure, mindfulness, and stress alleviation. Embrace the adventure of plant motherhood, learn from your experiences, and enjoy the pleasures of a growing indoor sanctuary.

In the following chapter, we'll get into the realities of plant care, including watering tactics, light needs, fertilization options, and typical insect and disease challenges. We'll empower you with the information and skills to solve any difficulties and guarantee your leafy companions flourish for years to come.

CHAPTER 7

Plant Parenthood 101: Essential Care for Indoor Plants

Now that you've welcomed your new green friends into your house, it's time to dig into the realities of plant maintenance. This chapter gives you with the crucial information and skills to guarantee your indoor sanctuary flourishes. We'll discuss lighting requirements, watering approaches, humidity demands, appropriate temperature ranges, and fertilising tactics.

Lighting: Illuminating the Path to Growth

Light is the lifeblood of plants. Understanding your plant's exact light needs is vital for its health and development. Here's a breakdown of typical light categories:

Bright, Indirect Light: This refers to filtered sunlight or well-lit spaces without direct sun exposure. Many popular houseplants, such as snake plants, spider plants, and peace lilies, thrive in this light spectrum.

Medium Light: This category comprises places with moderate sunshine exposure for a few hours a day. Dracaena, philodendrons, and ZZ plants are examples of plants that survive medium light conditions.

Low Light: Certain plants may survive and even flourish in poorly illuminated conditions. These include snake plants, pothos (Epipremnum aureum), and ZZ plants. However, low-light plants may suffer delayed development compared to those getting greater light.

Direct sunshine: Some plants, including cactus and succulents, need several hours of direct sunshine everyday. However, too direct sunshine may burn leaves on more sensitive plants.

Tips for Optimal Lighting:

Observe Natural Light Patterns: Throughout the day, pay attention to how sunshine flows around your living environment. Identify regions with bright, indirect light, medium light, and low-light zones.

Rotate Plants Regularly: For plants that prefer indirect light, rotate them occasionally to maintain uniform development on all sides.

Augment with Grow Lights (Optional): If natural light is inadequate, try utilising artificial grow lights to augment your plants' requirements. Choose lights with a spectrum ideal for plant development and modify the length and intensity depending on the unique needs of your plants.

Watering Wisdom: Avoiding the Overwatering Trap

Overwatering is a frequent factor behind plant failure. Different plants have varied watering demands, so knowing your plant's preferences is vital. Here are some broad guidelines:

The Finger Test: Stick your finger into the potting mix up to the first knuckle. If the soil seems dry, it's time to water. If it seems damp, wait a few days before checking again.

Observe the Leaves: Drooping leaves may be a symptom of thirst, but be cautious to rule out other reasons like inadequate light before watering.

Watering Frequency: Frequency will vary according on the plant, pot size, light conditions, and season. Generally, plants in brighter light and higher temps will need more regular watering compared to those in low-light, colder conditions.

Watering Techniques to Avoid Drowning Your Plants:

Water Thoroughly: When watering, attempt to soak the potting mix until water flows out the drainage holes. This ensures all the roots get moisture.

Empty Drainage Trays: Never let your plant to sit in water. Empty any water that gathers in the drainage tray after watering to avoid root rot.

Consider Bottom Watering: For certain plants, bottom watering might be a good method. Place the pot in a shallow dish filled with water. Allow the plant to soak the water from the bottom for approximately 30 minutes, then remove it from the dish.

Humidity Hacks: Keeping Your Plants Happy and Hydrated

Humidity refers to the quantity of moisture in the air. Many tropical plants demand moderate to high humidity levels, which may be tough to manage in dry indoor conditions, particularly during winter months. Here are several techniques to boost humidity around your plants:

Grouping Plants: Grouping plants together generates a microclimate with increased humidity levels.

Pebble Trays: Place a tray loaded with stones and water beneath your plant container. As the water evaporates, it will enhance humidity surrounding the plant. Be sure the pot rests above the waterline to avoid root rot.

Humidifier: Consider utilising a humidifier to boost general humidity levels in your house, especially good for big plant collections or those demanding high humidity.

Temperature Talk: Finding the Perfect Climate Indoors

Most houseplants enjoy mild temperatures between 65°F and 75°F (18°C and 24°C). Avoid putting plants near drafty windows, air vents, or heat sources, since these might produce temperature changes and stress the plant.

Feeding Frenzy: Fertilizing for Healthy Growth

Fertilization may give critical nutrients and encourage healthy development, particularly for plants with lush foliage or blooming kinds. However, overfertilizing may be hazardous, leading to salt accumulation in the soil and possible damage to the plant. Here's a summary of fertilization basics:

Types of Fertilizers: Most houseplant fertilizers come in liquid or water-soluble granular forms. Organic fertilizers, generated from natural sources, deliver nutrients slowly, whereas synthetic fertilizers give a rapid boost.

Dilution is Key: Always dilute fertilizer according to the manufacturer's directions. Using a lesser solution more often is typically safer than applying a high concentration occasionally.

Frequency Matters: Most houseplants only require fertilizer during their active growth season, often spring and summer. Reduce or halt fertiliser throughout autumn and winter when growth slows down.

Signs Your Plant Needs Fertilizer: Slow growth, pale leaves, or stunted new foliage might indicate a nutritional deficit. However, rule out other probable problems like inadequate light or underwatering before resorting to fertilization.

Additional Tips for Plant Care Success

Repotting: As your plant develops, it may need repotting into a bigger container. Signs like roots circling the container or the plant becoming top-heavy signal the need for repotting. Choose a pot just slightly bigger than the existing one to prevent overwatering. Use a new, well-draining potting mix before repotting.

Cleaning Leaves: Dust may collect on plant leaves, reducing their capacity to absorb light and photosynthesis. Wipe leaves lightly with a moist cloth frequently to eliminate dust and promote their general health.

Pest and Disease Control: Houseplants may be subject to pests like mealybugs, aphids, and spider mites. Early diagnosis and treatment are critical. Neem oil spray or insecticidal soap are great solutions for managing common pests. Fungal infections may also damage plants. Isolate problematic plants and treat them with fungicide according to the manufacturer's recommendations.

Observe and Adapt: Plants are living entities and their demands may alter over time. Pay attention to your plants and adapt their care appropriately. Observe for symptoms of stress such drooping leaves, leaf discolouration, or stunted development. Research probable reasons and change watering, lighting, or fertilization procedures as required.

The Joy of Plant Parenthood

Caring for indoor plants is a fascinating and satisfying experience. By following these important care rules, you can guarantee your green friends flourish and bring beauty, vitality, and well-being into your house. Remember, the adventure of plant motherhood is a constant learning process. Embrace the obstacles, appreciate the accomplishments, and enjoy the ever-evolving beauty of your indoor refuge.

CHAPTER 8

Troubleshooting Tips: Keeping Your Indoor Plants Thriving

Even with the greatest intentions, issues may develop in the realm of plant motherhood. Don't despair! This chapter gives you with the information and techniques to diagnose common plant problems, treat pest and disease concerns, revive wilting plants, and build a successful plant care routine.

Recognizing Common Plant Problems

The key to overcoming plant issues resides on early discovery. Here's an overview of several prevalent problems and their telltale signs:

Watering Issues:

Overwatering: Drooping leaves, yellowing foliage, and mushy stems may all suggest overwatering. Feel the soil - if it's damp, hold off on watering and let the soil to dry out fully before beginning.

Underwatering: Wilting foliage, dry and crispy leaves, and stunted development are indicators of underwatering. Water thoroughly when the top inch of soil seems dry.

Light Issues:

Insufficient Light: Stretched, leggy growth, pale leaves, and leaf drop may occur when plants don't get enough light. Move the plant to a brighter position or supplement with grow lights, if required.

Too Much Direct sunshine: Leaf burn, withering, and dry, crispy leaves indicate excessive sunshine exposure. Move the plant to a spot with indirect light.

Nutrient Deficiency:

Slow growth, pale leaves, or stunted new foliage might be symptoms of nutritional shortage. However, rule out other probable explanations before resorting to fertilization. If appropriate, fertilize throughout the growth season using a diluted solution according to the manufacturer's recommendations.

Pests and Diseases:

Look for symptoms of pests like mealybugs, aphids, or spider mites. These might appear as small insects on the leaves or stems, or leave behind sticky residue. Fungal illnesses may emerge as dark patches, wilting, or fuzzy mold on leaves.

Other Potential Issues:

Temperature extremes: Avoid planting plants near drafts, air vents, or heat sources.

Humidity problems: Dry indoor air may affect certain plants. Increase humidity using pebble trays, a humidifier, or placing plants together.

Repotting needs: Rootbound plants may exhibit indications of reduced growth or frequent wilting. Consider repotting into a slightly bigger container with new potting mix.

Dealing with Pests and Diseases

Early response is vital when dealing with pests and illnesses. Here are several strategies:

Isolation: Isolate any sick plants to avoid spreading the problem to others.

Natural Methods: For light infestations, try wiping leaves with insecticidal soap or neem oil spray. Be sure to follow application directions thoroughly.

Insecticidal Soap or Neem Oil Spray: These organic choices might be useful for managing common pests.

Fungicides: For fungal infections, follow the directions on a fungicide particularly labeled for houseplants.

Severe Cases: In extreme circumstances, you may need to remove the sick plant to avoid further spread.

Reviving Wilting Plants

A withering plant doesn't always indicate it's a lost cause. Here's what you can do:

Identify the Cause: Check the soil wetness. If underwatered, water thoroughly. If overwatered, let the soil to dry out entirely before starting watering.

enhance Humidity: Mist the plant or use a pebble tray to enhance humidity around it.

Adjust Light: Move the plant to a more optimal light position depending on its demands.

Prune Away Dead Leaves: Remove any dead or damaged leaves to stimulate fresh development.

Repotting for Continued Growth

As your plant develops, it may need repotting into a bigger container. Here are several symptoms that suggest repotting is necessary:

Roots round the pot: If roots are visible at the soil surface or growing out of drainage holes, the plant is likely rootbound and requires a bigger pot.

Frequent wilting: Even after watering, the plant wilts fast if the container is too tiny and doesn't retain enough moisture.

Stunted growth: The plant may cease developing or display stunted new leaves if it's rootbound.

Repotting Tips:

Choose the proper pot: Select a pot just slightly bigger than the present one. A pot that's too big might lead to overwatering difficulties.

Use fresh potting mix: Never reuse old potting mix, since it might hold pests and illnesses. Opt for a well-draining potting mix particularly intended for indoor plants.

Gently loosen the roots (continued): in the new pot. Avoid harming healthy roots.

Water thoroughly: After repotting, water the plant well to settle the dirt around the roots. Avoid overwatering in the following weeks while the plant establishes itself in the new container.

Creating a Plant Care Routine

Developing a regular plant care regimen is crucial to their long-term success. Here are some recommendations for building a routine:

Schedule Regular Watering: Depending on the plant and its surroundings, select a watering schedule and adhere to it as closely as feasible. Feel the soil periodically to adjust watering frequency as required.

Rotate Plants for equal development: For plants that prefer indirect light, rotate them occasionally to maintain equal development on all sides.

Examine Leaves Regularly: Take a time each week to examine your plants for indications of pests, illnesses, or other difficulties. Early identification allows for rapid response and avoids issues from worsening.

Fertilize During Growing Season: During spring and summer, fertilize your plants according to the manufacturer's directions and the individual demands of each plant. Hold off on fertiliser throughout autumn and winter when growth slows down.

Clean Leaves Periodically: Wipe dust off leaves with a moist cloth to boost their capacity to absorb light for photosynthesis.

Repot When Necessary: Pay attention to the indicators that suggest a plant requires repotting, and address it immediately to guarantee sustained healthy development.

Embrace the Learning Journey

Plant motherhood is a constant learning process. There will be accomplishments and difficulties along the road. Don't get disheartened by setbacks. Research answers to issues, alter your care techniques as appropriate, and celebrate your triumphs. With time and devotion, you'll gain a great awareness of your plants' requirements and be rewarded with a flourishing indoor retreat.

CHAPTER 9

Beyond the Basics: Styling Your Home with Indoor Plants

Having developed a foundation in plant care, it's time to upgrade your interior area with the smart placement and tasteful presentation of your green friends. This chapter digs into the art of indoor plant decorating, examining inventive methods to incorporate plants into your home for a coherent and visually pleasing setting.

Choosing the Right Pots and Planters

While the fundamental purpose of a pot is to house your plant and its root system, it also plays an important aesthetic role. Here's how to pick the correct pots and planters:

Material: Pots and planters come in a broad range of materials, including terracotta, ceramic, glazed pottery, plastic, and metal. Consider the overall design of your house and pick materials that compliment your décor. Terracotta pots give breathability, whereas glazed ceramic or plastic pots may retain moisture better.

Size: The pot size should be appropriate to the plant's maturity. A pot that's too big might lead to overwatering difficulties. Choose a pot that provides for some area for root development, but not overly so.

Color and Design: Pots and planters come in a multitude of colors and designs. Opt for neutral-colored pots for a minimalist design, or use vibrant colors and patterns to give a bit of individuality. Consider utilising containers that compliment the leaves or blossoms of your plants.

Drainage: Ensure your pots have drainage holes to avoid waterlogging, which may harm plant roots. If using decorative planters without drainage, put the plant inside a nursery pot with drainage holes and insert it into the decorative planter.

Grouping Plants for Maximum Impact

Grouping plants may create a visually appealing focal point and increase the sensation of lushness in your area. Here are some ideas for efficient grouping:

Size & Scale: Combine plants of varied heights and sizes for a layered look. Taller plants may provide as a background for smaller, trailing kinds.

Light needs: Group plants with comparable light needs together to ensure they all flourish in their allotted spot.

Foliage Color and Texture: Play with opposing or complimentary foliage colors and textures for a visually fascinating arrangement. For example, match a plant with variegated leaves next to one with plain green foliage, or blend plants with smooth leaves with others that have a fuzzy or textured feel.

Container Harmony: Use pots and planters in a same design or color palette when combining plants to produce a coherent effect.

Utilizing Vertical Space with Hanging Plants

Limited floor space doesn't have to limit your indoor jungle aspirations. Hanging plants provide a bit of whimsy and use vertical space wisely. Here's how to include them:

Choosing the Right Plants: Trailing kinds like pothos (Epipremnum aureum), spider plants (Chlorophytum comosum), and philodendrons are perfect for hanging baskets. Consider the weight of the mature plant while picking a hanger.

Placement: Hang planters in locations with ample indirect light. Avoid hanging plants directly over furniture or pathways to avoid them from blocking movement.

Watering Considerations: Hanging plants may dry out quicker owing to increased air movement. Check the soil moisture periodically and water when the top inch seems dry.

Creating a Living Wall Oasis

Living walls turn a vertical surface into a bright and air-purifying green environment. Here's things to consider before making a living wall:

Space and Support: Living walls need robust support and appropriate space. Ensure the wall can take the weight of the plants and their growth media.
Lighting: Choose an area that gets adequate natural light for the plants you wish to employ.
Watering System: Living walls frequently need sophisticated irrigation systems to maintain uniform watering and avoid overwatering or underwatering. Explore pre-made living wall systems or talk with a specialist for unique options.

Plant Selection: Opt for low-maintenance plants that sustain moderate moisture levels and don't need regular repotting. Ferns, mosses, and bromeliads are common alternatives for living walls.

Indoor Plant Design Inspiration

Here are some design ideas to ignite your imagination and encourage you to integrate plants into your home:

Bookshelves and Built-in Units: Showcase trailing plants or succulents on shelves to lend a touch of life to bookshelves or built-in units.
Coffee and Side Tables: Adorn coffee tables and side tables with tiny potted plants to generate a feeling of liveliness.
Windowsills: Fill sunny window sills with light-loving plants like herbs or cactus to liven up the room.
Bathroom Retreat: Introduce moisture-loving plants like ferns or peace lilies to your bathroom to create a spa-like feel.

vacant Corners : Fill vacant corners with a tall floor plant like a Fiddle Leaf Fig (Ficus lyrata) or a Monstera Deliciosa for a dramatic statement piece.

Kitchen Inspiration: Hang herb plants near a window for quick access to fresh herbs and a splash of greenery in your kitchen room.

Dining Table Centerpiece: Create a lively centerpiece for your dining table with a low-growing arrangement of succulents or a terrarium packed with small plants.

Bedroom Oasis: Opt for low-light tolerant plants like Snake Plants (Sansevieria Trifasciata) or ZZ Plants (Zamioculcas zamiifolia) in your bedroom to encourage relaxation and enhance air quality.

Remember: When designing your house with plants, emphasise their well-being. Choose settings with acceptable light conditions and ensure they get sufficient maintenance. A healthy plant is not only visually beautiful but also helps to a healthier interior atmosphere.

Beyond Aesthetics: The Functional Benefits of Indoor Plants

While the visual attraction of indoor plants is clear, their advantages extend well beyond aesthetics. Here are some reasons to embrace houseplants:

Air Purification: Plants naturally filter toxins and pollutants from the air, improving indoor air quality and boosting respiratory wellness.

Stress Reduction: Studies have shown that engaging with plants may lower stress levels and enhance emotions of relaxation and well-being.

Increased Productivity: The presence of plants at an office may boost focus, creativity, and overall productivity.

Boosts Mood: Caring for plants may be a satisfying and relaxing activity, generating a feeling of achievement and enhancing mood.

Creates a Sense of Connection with Nature: Indoor plants bring a touch of the natural world within, encouraging a sense of connection with nature and generating a feeling of calm and serenity.

Embrace the Journey of Indoor Plant Styling

There's no one "right" approach to design your house with plants. Experiment, have fun, and let your creativity flow. As your plant collection increases, so too will your knowledge and confidence. Let your indoor paradise develop with your interests and preferences, and enjoy the continual process of building a lively and life-enhancing area.

CHAPTER 10

The Green Revolution: Indoor Plants for a Sustainable Future

Throughout this book, we've covered the delights and advantages of incorporating indoor plants into your life. But the good influence goes well beyond personal well-being. Plants play a critical part in building a more sustainable future, and adopting houseplants is a simple but essential step towards a greener society. This chapter digs into the environmental advantages of indoor plants, their contribution to better air quality and energy efficiency, and their capacity to foster a closer connection with nature.

The Environmental Benefits of Indoor Plants

Plants are the quiet heroes of our world, working as the lungs of the Earth. Here's how bringing them inside leads to a more sustainable future:

Air Purification: Plants operate as natural air filters, collecting carbon dioxide and releasing oxygen. This procedure helps eliminate typical indoor air pollutants including formaldehyde, benzene, and trichloroethylene, which may be detrimental to human health. By increasing indoor air quality, plants help alleviate respiratory difficulties and allergies.

Combating Climate Change: Plants take carbon dioxide, a significant greenhouse gas, and turn it into oxygen via photosynthesis. This mechanism plays a key role in mitigating climate change by decreasing greenhouse gas levels in the atmosphere. While indoor plants have a lower influence compared to outside forests, every little contributes.

Reduced Energy Consumption: Strategically planted plants may offer shade and cooling benefits, especially during hot summer months. This may assist minimise energy usage by lowering dependency on air conditioning systems.

Sustainable Materials: Many popular houseplants may be reproduced by cuttings or division, lowering the need to buy new plants and reducing the environmental impact associated with mass manufacturing.

Improving Indoor Air Quality and Energy Efficiency

In today's society, we spend a substantial amount of time inside, where air quality may be damaged by toxins from construction materials, cleaning chemicals, and even furniture.

Houseplants provide a natural approach to enhance indoor air quality:

Choosing the Right Plants: Certain plants are especially good at eliminating certain contaminants. Research variants like Snake Plants (Sansevieria Trifasciata), Spider Plants (Chlorophytum comosum), and Peace Lilies (Spathiphyllum wallisii) for their air-purifying characteristics.

Plant Placement: For best air filtration, place multiple plants together in areas where you spend the most time. Placing plants near windows helps them to get sunshine, vital for their air-purifying function.

Beyond Plants: While plants clearly contribute to better air quality, they are not a comprehensive solution. Proper ventilation procedures remain vital. Consider employing air purifiers for further filtering, particularly in heavily crowded settings.

Cultivating a Connection with Nature

In an increasingly urbanized society, our relationship with nature might become strained. Houseplants provide an easy method to bring a bit of the outdoor world indoors:

Biophilia and Well-being: Biophilia refers to our intrinsic human connection with nature. Studies have shown that connecting with plants may lower stress levels, increase mood, and promote general well-being.

Mindfulness and Appreciation: Caring for plants demands us to slow down, observe, and enjoy the natural environment. This promotes a feeling of awareness and cultivates a deeper connection with the environment.

Inspiring Sustainability: The beauty and resilience of plants may encourage us to embrace more sustainable habits in our everyday lives. Indoor plant parenting may be a starting stone towards a more mindful and ecologically sustainable lifestyle.

Embrace the Power of Indoor Plants

By bringing indoor plants into your life, you're not only creating a beautiful and healthy living place; you're contributing to a greener future. Plants provide a number of advantages, from increasing air quality and energy efficiency to developing a closer connection with nature. Every plant you nurture helps to the collective effort towards a more sustainable world.

The Green Revolution Starts at Home

The green revolution doesn't need spectacular gestures. It begins with basic decisions in your own house. By adopting houseplants, you join a worldwide movement devoted to environmental well-being. Share your enthusiasm with friends and family, encourage others to develop their own green spaces, and together, we can build a healthier world, one plant at a time.

Final Words: A Journey of Growth and Discovery

This book has hopefully provided you with the information and resources to begin on your adventure of indoor plant motherhood. Remember, the path of plant care is a constant learning process. Don't be disheartened by setbacks, embrace them as chances to learn and improve. As your plant collection expands, so too will your knowledge and confidence. Enjoy the process of building a healthy indoor sanctuary, enjoy in the beauty and delight that plants provide, and celebrate the beneficial influence you're having on your personal well-being and the environment.

CONCLUSION

Throughout this book, we've dug into the fascinating world of indoor plants. We've examined the science behind their advantages, from air purification to stress relief, and found the enchantment they bring to our lives. We've learnt about the necessary care techniques for many sorts of plants, arming you with the knowledge to cultivate a healthy indoor sanctuary. But the influence of houseplants goes well beyond creating a lovely atmosphere.

The Power of Plants in Our Lives

Plants are more than simply cosmetic components. They are living creatures that provide a plethora of advantages for our well-being and the environment. Here's a closer look at the power plants wield in our lives:

Enhanced Physical Health: Studies have shown that exposure to plants may enhance lung health by decreasing contaminants in the air. Additionally, plants may assist manage humidity levels, which can be good for individuals prone to allergies or respiratory ailments.

Mental and Emotional Well-being: Caring for plants may be a source of stress reduction and relaxation. Studies show that engaging with plants may reduce cortisol levels, the stress hormone, and boost feelings of relaxation and well-being. The act of caring another living creature and observing its progress may promote a feeling of achievement and increase overall happiness.

Increased Focus and Productivity: Research suggests that the presence of plants in a workstation may promote cognitive performance, improve focus, and stimulate creativity.

Connection with Nature: In an increasingly urbanized world, houseplants provide a means to bring a bit of nature inside. This connection with the natural world develops a feeling of biophilia, our fundamental human love of nature, and has been demonstrated to lead to enhanced mental and emotional well-being.

A Gateway to Mindfulness: Caring for plants encourages us to slow down, notice, and enjoy the natural environment. The process of watering, trimming, and just watching our green friends develops attention and a deeper connection with the present moment.

A Greener Future Starts at Home

The environmental advantages of indoor plants are as compelling. By adding plants into your life, you can create a beneficial influence on the planet:

Improved Indoor Air Quality: Plants operate as natural air filters, eliminating hazardous pollutants from the air and increasing overall indoor air quality. This is especially advantageous in newer structures with tight sealing and little ventilation.

Combating Climate Change: Plants absorb carbon dioxide, a significant greenhouse gas, and release oxygen via photosynthesis. While indoor plants have a modest influence compared to large forests, every plant helps to the worldwide effort to combat climate change.

Reduced Energy Consumption: Strategically planted plants may offer shade and cooling benefits, especially during hot summer months. This may assist minimise energy usage by lowering dependency on air conditioning systems.

Sustainable Practices: Many houseplants may be grown from cuttings or division, lessening the need for regular purchases from commercial growers and lowering the environmental imprint associated with mass manufacturing.

Beyond the Personal: A Collective Effort

Embracing houseplants is not a single endeavour. By sharing your love with friends and family and motivating others to develop their own green spaces, we can build a collective movement towards a healthier world, one plant at a time. Here are some ways to spread the green love:

Present the Joy of Plants: Sharing a plant with a friend or loved one is a meaningful and sustainable present that goes on giving.

Spread the Word: Talk to others about the advantages of indoor plants and inspire them to appreciate their own green hideaway.

Support Local Growers: Patronize local nurseries and plant stores that encourage sustainable methods.

Get Involved: Join online groups of plant enthusiasts, share your experiences, and learn from others.

Embrace the Journey, Celebrate the Growth

The path of plant motherhood is one of ongoing learning and discovery. There will be accomplishments and difficulties along the road. Embrace the learning opportunities given by failures, and appreciate your wins, large or little. As you maintain your plants, you'll build not just a healthy indoor paradise but also a stronger connection with nature, a feeling of well-being, and a good influence on the environment.

This book serves as a beginning point for your trip. As you continue to explore the world of indoor plants, you'll find numerous new types, perfect your care practices, and create your own distinct green thumb. Embrace the process, appreciate the enchantment, and revel in the "Green Cure" that houseplants provide for a happier, healthier you and a more sustainable future for our world.

Additional Resources

Books and Online Resources (continued): Join online forums devoted to houseplants to interact with other enthusiasts, exchange experiences, and learn from one another.

Nurseries and Plant Shops: Local nurseries and plant shops provide a treasure trove of plant types, professional guidance, and hands-on learning opportunities. Support companies that promote sustainable processes and ethical procurement of plants.

seminars and Events: Many botanical gardens, nurseries, and community centres provide seminars and events focusing on plant care, propagation methods, and particular plant groupings. These events give great learning opportunities and an opportunity to network with other plant enthusiasts.

The Final Bloom

May your adventure with indoor plants be one of ongoing development and discovery. Let your house be a refuge full with bright vitality, and may your connection with the natural world increase with each blooming leaf. Remember, the tiniest deeds, when multiplied by many, may produce a huge effect. So, maintain your plants, share your enthusiasm, and together, let's grow a greener future, one leaf at a time.

APPENDIX

Your voyage into the world of indoor plants has just started! This appendix gives you with important resources to enhance your knowledge, interact with other plant lovers, and guarantee your green friends flourish. Explore the following resources to develop your plant parenting abilities and build a healthy indoor retreat.

Plant Identification Guides

Identifying your plant is the first step to giving adequate care. Here are some useful resources for correct plant identification:

smartphone applications: Several user-friendly smartphone applications can assist you identify plants based on photographs. Popular choices include:

PlantSnap: This app enables you to snap a photo of your plant and obtain fast identification ideas.

LeafSnap: Similar to PlantSnap, LeafSnap utilises image recognition technology to identify plants based on images.

Seek by iNaturalist: This thorough app not only helps identify plants but also links you to a worldwide network of naturalists who can contribute extra information.

Online Plant Identification Websites: These websites provide huge databases of plant photographs and thorough descriptions to help with identification.

The Spruce: https://www.thespruce.com/ provides a wide variety of plant profiles with high-quality images and maintenance instructions. Their "Identify a Plant" option enables you to submit an image for possible identification.

Garden Guides: https://www.gardenguides.com/ provides a user-friendly plant identification tool with search filters based on numerous plant traits.

The RHS Plant Finder: Developed by the Royal Horticultural Society, this website gives thorough information on a broad selection of plants, including identification hints and care advice. (https://www.rhs.org.uk/plants)

Books: Invest in an excellent quality plant identification manual suited to your location. These publications often offer photographs, descriptions, and care needs for a range of houseplants.

Online Plant Care Communities

The internet provides a plethora of knowledge and a dynamic community of plant lovers. Here are some useful internet tools to connect with individuals that share your passion:

Online Forums and Discussion Boards: Several online forums and discussion boards devoted to houseplants give a platform to ask questions, exchange experiences, and learn from other plant parents. Popular choices include:

Reddit Houseplants: This vast and active subreddit provides a plethora of knowledge on all things houseplants. Users may submit inquiries, share images of their plants, and join in conversations on many themes. (https://www.reddit.com/r/houseplants/)

Facebook Groups: Search for Facebook groups devoted to indoor plants in your region or with specialised hobbies like succulents, cactus, or air plants. These communities provide a helpful network for exchanging insights, fixing difficulties, and finding inspiration.

Online Gardening Communities: Gardening websites like https://www.amazon.com/The-Garden/dp/B0B6D6N42H and [https://garden.org/](htttps://garden.org/) typically offer active forums where you can interact with other plant aficionados.

Social Media: Follow social media profiles of plant influencers, nurseries, and botanical gardens for daily inspiration, plant care advice, and new plant discoveries. Platforms like Instagram and Pinterest provide a visual feast of stunning houseplants and imaginative display ideas.

YouTube Channels: Several instructive YouTube channels include plant care lessons, plant evaluations, and behind-the-scenes insights inside greenhouses and nurseries. Explore channels like Planterina, Crazy Plant Lady, and Plantaholic for great information and a bit of fun.

Local Nurseries and Garden Centers

Local nurseries and garden stores are a treasure mine of information for plant aficionados. Here's how to optimise your experience:

Expert Advice: Nursery workers are frequently informed plant specialists who can answer your concerns about particular plant care, detect possible issues, and propose appropriate plants for your requirements. Don't hesitate to seek their guidance and experience.

seminars and Events: Many nurseries and garden centers provide seminars and events focusing on plant care, propagation methods, and particular plant groupings. These events give important learning opportunities and an opportunity to network with other plant enthusiasts in your town.

Plant collection: Explore the broad collection of plants available at local nurseries. Look for healthy specimens with vivid foliage and no evidence of pests or illnesses. Purchasing from local shops provides fresher plants and helps your neighbourhood.

Sustainable Practices: Support nurseries that promote sustainable practices, such as employing organic pest management techniques and getting plants from trustworthy farmers.

Additional Resources

Botanical Gardens: Botanical gardens provide a beautiful and informative experience. Explore the vast plant collections, learn about numerous plant species, and attend seminars or lectures provided by botanical garden personnel.

Libraries: Local libraries generally contain a multitude of information on houseplants, including books, publications, and online databases. Utilize these resources to increase your knowledge and discover new plant kinds.

Cooperative Extension programmes: Many universities and colleges have cooperative extension programmes that provide gardening guidance and instructional materials. These services may be a great source of information on plant care, pest control, and sustainable gardening methods. Check your local university extension page for information.

Plant Propagation Resources: If you're interested in propagating your own plants, browse internet resources and publications devoted to this subject. Learn about diverse propagation methods including division, cuttings, and grafting, and build your plant collection without breaking the bank.

Remember: The secret to success in the world of indoor plants is continual learning and a willingness to try. Don't be hesitant to ask questions, seek guidance, and accept the obstacles along the path. The advantages of a flourishing indoor paradise and the thrill of fostering life are well worth the effort.

Happy Planting!

www.ingramcontent.com/pod-product-compliance
Lightning Source LLC
Chambersburg PA
CBHW050110230526
45470CB00004B/1758